Putting the World on a Page

by Sarah Mastrianni

Photographs by Russell Pickering

Developed for Harcourt, Inc., by Gareth Stevens, Inc. This edition published by Harcourt, Inc., by agreement with Gareth Stevens, Inc. No part of this publication may be reproduced or transmitted in any form or by any means, electronic or mechanical, including photocopy, recording, or any information storage and retrieval system, without permission in writing from the copyright holder.

Requests for permission to make copies of any part of the work should be addressed to Permissions Department, Gareth Stevens, Inc., 330 West Olive Street, Suite 100, Milwaukee, Wisconsin 53212. Fax: 414-332-3567.

HARCOURT and the Harcourt Logo are trademarks of Harcourt, Inc., registered in the United States of America and/or other jurisdictions.

Printed in the United States of America

ISBN-13: 978-0-15-360189-7
ISBN-10: 0-15-360189-2

1 2 3 4 5 6 7 8 9 10 175 16 15 14 13 12 11 10 09 08 07

Harcourt
SCHOOL PUBLISHERS

Chapter 1:

Stamping Through Australia and Japan

Julia's grandfather is a philatelist. A philatelist is a person who collects postage stamps. During his travels, Grandfather finds and purchases stamps. He then saves them as souvenirs. When he receives letters from family and friends, he often keeps the stamps from the envelopes, too. Over the years, he has collected many postage stamps that way.

Today, Julia's grandfather wants Julia to help him organize all of his stamps into one large stamp album. Julia is excited about this special project. She enjoys spending time with Grandfather. He tells fascinating stories about how each stamp came to be part of his collection. Julia and Grandfather plan how they will organize the new stamp album.

Grandfather has a large collection of stamp albums. He arranged his original stamp albums by country, one album for each country. Over the years, the albums have become disorganized. Now, many stamps are out of place, and are no longer located with other stamps from the same country.

Julia thinks arranging the stamps by country is a fabulous idea. First, Julia and Grandfather will sort the stamps by country. Then, they will place the stamps in the new album for that country. Julia flips through Grandfather's stamp albums. She finds some interesting stamps from Japan. That's a good place to start. Julia begins to arrange the first page of the new stamp album. She makes nine rows of four stamps each.

These stamps are from Japan.

These Australian stamps include birds, flowers, and people.

While Julia arranges stamps from Japan, Grandfather searches through his albums for stamps from Australia. He locates several interesting ones. In fact, he finds enough stamps to make six rows of six stamps. Julia looks at the variety of pictures on the stamps. She notices images of birds, flowers, and people of Australia.

Grandfather points to one stamp in particular. This stamp features a huge white whale off the coast of Australia. Grandfather knows that Julia loves whales so she will love this stamp. He tells Julia that at one time Australia issued a series of stamps called "Creatures of Slime." Julia laughs. She tries to imagine what slimy animals would be pictured on those stamps.

Julia and Grandfather pause a moment to look at their new page layouts. They need to decide the best way to display the many international stamps throughout the new album. They use multiplication. This allows them to find out how many stamps are on the pages. Julia multiplies 9 × 4. Grandfather multiplies 6 × 6.

Julia and Grandfather smile. They have different layouts, but the same number of postage stamps on their pages. They agree that 36 must be the perfect number of stamps for a page. They also agree to use Julia's page layout of nine rows of four stamps each in the new album.

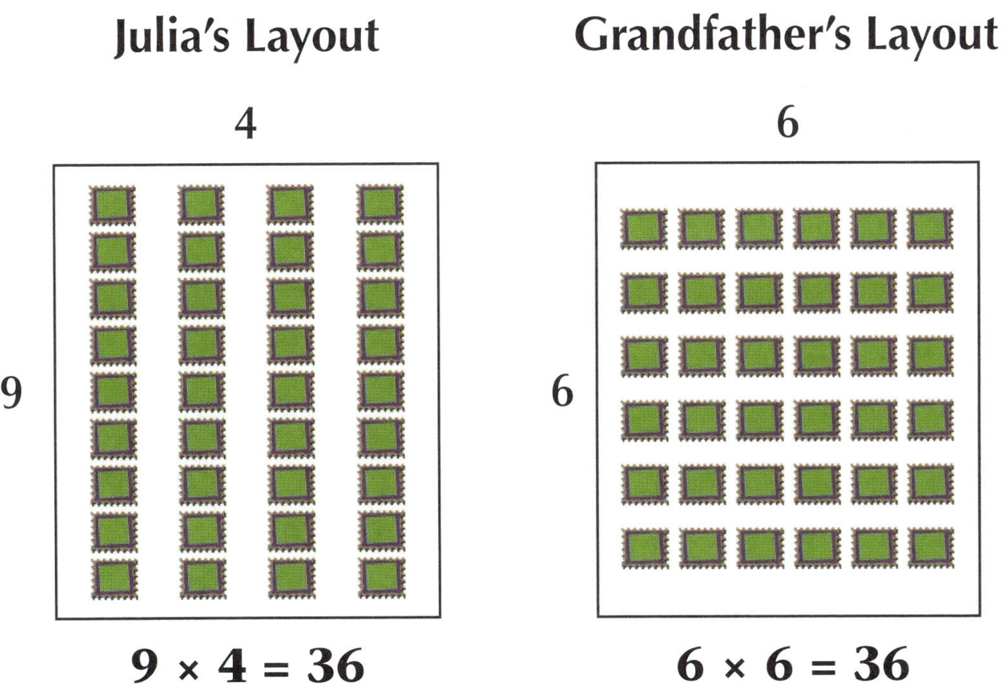

Chapter 2:

Stories from Stamps

Grandfather's collection contains stamps from many countries. In fact, Grandfather thinks that the stamps in his collection represent 18 different countries. Some of his favorite stamps are from the United States. Grandfather finds a different page that has many stamps from the 1996 Summer Olympic Games, which were held in Atlanta, Georgia. Grandfather is especially fond of these stamps because he and Grandmother actually attended some of the Olympic events.

Grandfather tells wonderful stories about his experiences and the stamps he has collected. His stories make Julia wish that she were with her grandparents on their travels. Julia is eager to hear stories about the other stamps, too.

Grandfather continues to share his memories of the 1996 Summer Olympics while Julia looks at the stamps. He tells about races won and the efforts of the athletes at the Olympics. Soon they get back to organizing the new stamp album. Grandfather thinks there are enough stamps that they will fill two pages of the album for each country. Julia wonders if there are enough empty pages to hold all those stamps. Julia uses multiplication to find out.

$$18 \times 2 = 36$$

There are postage stamps from 18 countries. Each country will fill two pages in the album. They need 36 pages. The new stamp album has 40 pages. It will be the perfect size.

This photo album has enough pages for Grandfather's stamp collection.

This stamp commemorates the Mars Rover Sojourner mission.

Julia places stamps from the United States onto a page. She finds a stamp that says "Mars Rover Sojourner." Julia is curious. She learned about Mars in science class, but she doesn't know about the Mars Rover Sojourner. Grandfather says that the Sojourner was a robotic vehicle sent to Mars in 1996 on the Pathfinder. The Pathfinder was an unmanned spacecraft.

Sojourner's mission was to collect data about the planet's rocks, minerals, and soil. Another part of the mission was to take thousands of photographs of Mars. This stamp was issued to commemorate the mission. Julia learns many interesting facts from her Grandfather's stories.

Julia figures out how long the project will take.

Julia thought that organizing the postage stamps into the new album would be a fast project at first. She is quite surprised at how much time it is taking. She looks at the clock and already four hours have passed! Julia and Grandfather have completed two pages each. They have completed four pages altogether.

Julia thinks for a moment and she figures out that it takes two hours to fill each page with the stamps. With 36 pages to complete, the project will take about 72 hours. Julia does not mind at all because she enjoys looking at Grandfather's stamp collection. She loves spending time with Grandfather. They are having fun together.

Chapter 3:

Julia: The Newest Philatelist

Shortly, Julia and Grandfather decide it is time for a little break. They share a healthy snack of vegetables and dip. As they eat, they talk about their project. Julia notices a stamp in Grandfather's collection that looks familiar. She points to the stamp. The stamp is similar to a painting hanging in her grandparents' home.

Grandfather smiles and explains that the picture on the stamp is a painting called "Big Raven." The painter, Emily Carr, is a well-known Canadian artist and writer who lived from 1871-1945. The stamp celebrates her life's work and was issued long after she died. Since Grandfather and Grandmother like Emily Carr's artwork, they bought one of her paintings years ago.

Julia wonders if some of the people, places, and events that she likes are pictured on stamps, too. She asks Grandfather that question. Grandfather thinks that there are many stamps with images Julia would recognize. Some might be in Grandfather's collection, but countries issue new stamps each year. Postage stamps might include people, places, and things Julia knows or recognizes.

Grandfather suggests that Julia do some research to find out more about the stamps that countries issue. That sounds like a good idea to Julia, so she decides that will be her next project. In fact, after she helps Grandfather organize his stamps, Julia wants to become a philatelist and start a stamp collection of her own.

Grandfather and Julia talk about stamps with familiar images.

Grandfather and Julia decide how to finish the album pages.

After their snacks, Julia and Grandfather decide not to return to organizing the stamp collection together. Instead, they decide to split up the remaining work. They will each work on the project separately during the week and they will meet again next weekend. At that time, they will see how much they have each accomplished.

Grandfather has a busy week planned so he will not have much time to work on the project. Julia is enjoying the project so much that she gladly agrees to complete a few extra pages. Before they clean up for the day, they decide how to share the remaining pages so they can finish the new stamp album.

There are stamps from 18 countries in Grandfather's collection. They completed album pages for two countries, Japan and Australia. They have some stamps from the United States in the album, too. There are stamps from 16 countries left to place in the album. Julia will complete album pages for ten countries. Grandfather will complete album pages for the remaining six countries.

Julia gives Grandfather the album pages he will need. Grandfather needs 12 album pages. That is enough for two pages for each of six countries. Julia needs 20 pages to finish her part of the project. That's enough for two pages for each of ten countries.

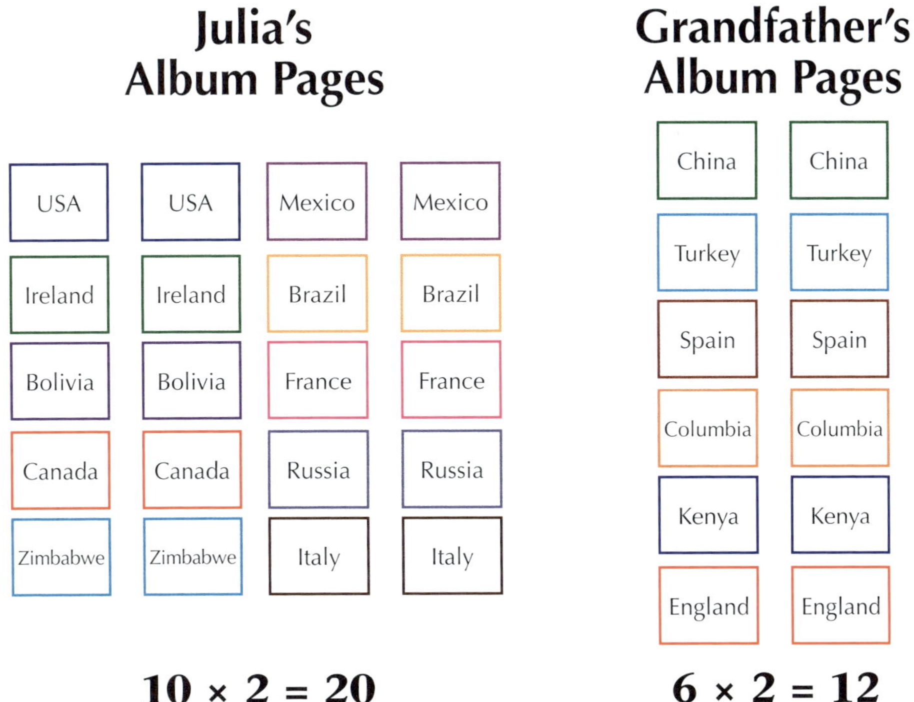

Before he leaves, Grandfather reminds Julia to handle the stamps carefully.

Grandfather reminds Julia to carefully handle the stamps. "Not all of the stamps are valuable," Grandfather says. "But, many of them have sentimental worth to me." Julia understands. She knows that many of the stamps remind Grandfather of special moments, people, and places that mean a lot to him. Julia promises to handle the stamps carefully, always using the stamp tongs.

Julia wonders if she completes 20 album pages and each page holds 36 stamps, how many stamps will she place in the album this week? She multiplies 20×36.

The math adds up to 720 stamps in all. Julia can't wait to look at all of the postage stamps. When she meets with Grandfather next weekend, she knows she will hear him tell some more interesting stories.

Julia gives Grandfather a hug as they say good-bye. She thanks him for a very special day. Just then, Grandfather reaches into his pocket and hands Julia a postcard. On the postcard is a beautiful stamp from Greece.

"This is for you," Grandfather says. "There is an interesting little story behind this stamp, too. That will have to wait until the next time we get together. Until then, keep this stamp in a safe place."

Julia smiles. She has the first stamp for her collection now. She can hardly wait to place it in her own stamp album. Soon, Grandfather will no longer be the only philatelist in the family!

This postcard from Greece includes the first stamp for Julia's collection.

Glossary

album a book with blank pages used to store things that are collected

arrange to place in order

commemorate to honor or remember

layout an arrangement or plan

philatelist a person who collects stamps

souvenir a small item that is a reminder of a place or an occasion